Which EV Truck Is Right for You?

A Consumer's Guide to Tesla Cybertruck, Rivian, Ford, and Hummer EVs

By

Jonathan Berkeley

TABLE OF CONTENTS

The Electric Vehicle Revolution

The turn of the century heralded a new era for the automotive industry, one defined by a paradigm shift toward sustainable and environmentally friendly transportation. Electric vehicles (EVs) have been at the forefront of this transformative journey, challenging old concepts of transportation and paving the way for a brighter future. We will delve into the key drivers, milestones, challenges, and overarching influence of the Electric Vehicle Revolution in this exploration. Concerns about climate change, air pollution, and the finite nature of fossil fuels began to take front stage around the turn of the century. The transportation sector, which is heavily reliant on internal combustion engines, contributed significantly to greenhouse gas emissions. As the globe grappled with the ramifications of environmental degradation, the need for a more sustainable mode of transportation became increasingly clear.

The origins of the Electric Vehicle Revolution may be traced back to the early 19th century, but it wasn't until the late 20th and early 21st centuries that electric vehicles gained widespread attention. The unveiling of the Tesla Roadster in 2008 was a watershed moment, demonstrating that electric automobiles could be both high-performance and aesthetically appealing. Tesla's success catalyzed intrest and investment in electric vehicle technology, pushing other automakers to join the race to electrification.

Advancements in battery technology are crucial to the success of electric vehicles. Over the last two centuries, there have been remarkable breakthroughs in battery chemistry, energy density, and production efficiency. These developments have not only expanded the range of electric vehicles, but have also made them more economically viable. Lithium-ion batteries in particular have become the industry standard, allowing EVs to compete with their internal combustion counterparts in terms of range and performance.

Government initiatives and policies aimed at encouraging sustainable mobility have had a huge impact on the electric vehicle revolution. Several countries have enacted incentives such as tax credits, rebates, and infrastructure investments to encourage the use of electric vehicles. These metrics have played a critical role in molding consumer behavior and creating a more favorable environment for the electric vehicle market to thrive.

One of the early impediments to the widespread adoption of electric vehicles was the lack of a comprehensive charging infrastructure. However, as the Electric Vehicle Revolution gained momentum, large investments were made to develop charging networks across the world. The development of fast-charging stations and the incorporation of charging infrastructure into urban design have reduced "range anxiety" and made electric vehicles more feasible for everyday use.

The Electric Vehicle Revolution is not limited to a single type of vehicle. It has given rise to a diverse range of electric vehicles, including sedans, SUVs, trucks, and even electric bicycles. This diversification of the electric vehicle market caters to a wide range of consumer preferences and needs, making sustainable mobility more accessible to a wider demographic.

The Electric Vehicle Revolution is fueled by a commitment to environmental sustainability. Electric vehicles emit zero tailpipe emissions, reducing air pollution and lowering transportation's influence on climate change. Furthermore, as the energy grid transitions to renewable sources, the overall carbon footprint of electric vehicles continues to decease, cementing their role as a key player in the search of a sustainable future. While the Electric Vehicle Revolution has made great strides, it is not without its difficulties. Issues like as battery disposal, resource extraction for battery manufacture, and the need for further technological advancements remain challenges. However, continued research, development, and collaborative efforts between governments, industries, and environmental organizations suggest a promising future for electric vehicles.

The Electric Vehicle Revolution has evolved from a niche concept to a driving force in the automotive sector during the last two decades. The incorporation of cutting-edge technology, supportive government policies, and a growing awareness of environmental issues have propelled electric vehicles into the mainstream. As we approach a new era in transportation, the Electric Vehicle Revolution serves as a beacon of hope, providing a blueprint to a more sustainable tomorrow.

The Significance of Electric Pickup Trucks in the Automotive Landscape

With the rise of electric vehicles (EVs), the automotive sector is experiencing a seismic upheaval, and electric pickup trucks have emerged as formidable players. Pickup trucks, traditionally associated with power, versatility, and a distinct American identity, have now entered a new era defined by sustainability and technological innovation. We will delve into the significance of electric pickup trucks in this exploration, examining their influence on the automotive landscape, consumer preferences, and the broader context of a world shifting to greener transportation alternatives.

Pickup trucks, traditionally associated with ruggedness and power, have long been a fixture on American roadways. Historically, these vehicles were powered by internal combustion engines that guzzled gasoline or diesel. The transition to electric power challenges the pickup truck segment's conventions, providing an alternative that accords with the worldwide movement toward sustainability. This break from tradition is a big step forward in redefining the identity of pickup trucks.

The environmental effect of classic internal combustion engine vehicles, particularly pickup trucks noted for their low fuel efficiency, is a developing concern. Electric pickup trucks address this issue head-on by eliminating tailpipe emissions, resulting in cleaner air and lower greenhouse gas emissions. As governments throughout the globe intensify their efforts to combat climate change, the adoption of electric pickup trucks becomes a critical component in achieving environmental sustainability targets.

While electric automobiles have gained popularity in urban settings, the debut of electric pickup trucks opens up new avenues in the automotive market. Pickup trucks, which are frequently associated with rural and suburban lifestyles, appeal to a demographic that may not have previously considered electric vehicles. This expansion of the electric vehicle audience is crucial for the wider adoption of sustainable mobility because it bridges the gap between traditional automotive preferences and the green revolution.

Tesla's Cybertruck's unveiling in 2019 catapulted electric pickup trucks into the worldwide limelight. The Cybertruck became a symbol of innovation and a catalyst for change in the pickup truck segment due to its unconventional design, robust specifications, and Elon Musk's ambitious aim. Aside from its striking appearance, the Cybertruck exemplifies the marriage of electric power with the rugged capabilities expected from a pickup, challenging industry standards and motivating competitors to lift their game.

Tesla's Cybertruck may have stolen the early focus, but it is not alone in the arena of electric pickup trucks. Rivian's R1T, Ford's F-150 Lightning, and GMC's Hummer EV truck are among the notable competitors in this brisk market. Whether it's Rivian's emphasis on adventure and off-road capabilities,

Ford's integration of the legendary F-150 brand, or GMC's reimagining of the Hummer as a electric powerhouse, each model has its own unique strengths. The competitive landscape not only encourages innovation, but also provides consumers with a diverse range of electric pickup alternatives.

Electric pickup trucks not only promise to be environmentally friendly, but they also offer a number of performance benefits. Electric motors provide fast torque, allowing for quick acceleration and superior towing capabilities. Furthermore, the lower center of gravity achieved by the placement of batteries contributes to increased stability, especially in off-road scenarios. These performance benefits call into question the concept that electric vehicles sacrifice power or utility.

While the benefits of electric pickup trucks are numerous, addressing consumer concerns is crucial. Range anxiety, or the fear of running out of battery power, has hampered widespread acceptance of electric vehicles. As a result, charging infrastructure plays a critical role in alleviating these concerns. As electric pickup trucks become more popular, the development of a comprehensive and efficient charging network is critical to ensuring their practicality for everyday use and long-distance travel.

The economic landscape surrounding electric vehicles is changing, with an emphasis on making them more accessible and cost-effective to consumers. While the initial purchase price of electric pickup trucks may be more than that of their traditional counterparts, considerations such as lower operating costs, reduced maintenance requirements, and potential government incentives all contribute to a compelling economic case for consumers. As technology advances and scale economies take effect, the cost of electric pickup trucks is expected to fall.

The shift to electric vehicles goes beyond the vehicles themselves and encompasses a broader ecosystem. As the demand for electric pickup trucks develops, so does the demand for skilled workers in manufacturing, research and development, and the establishment of charging infrastructure. This transformation not only has an influence on conventional automotive employment markets, but it also creates opportunities for new skill sets and industries to thrive.

Looking ahead, the importance of electric pickup trucks goes beyond environmental and economic ramifications. The integration of autonomous driving and connected features represents the future mobility direction. Electric pickup trucks outfitted with advanced driver-assist technologies contribute to the continuous evolution of transportation toward safer, more efficient, and interconnected systems.

While the electric pickup truck market is poised for expansion, challenges remain. The need for further advancements in energy density, charging speed, and materials remains a focus area for battery technology. Furthermore, the development of a comprehensive charging infrastructure, especially in rural and remote areas, is critical for ensuring the widespread adoption of electric pickup trucks.

The significance of electric pickup trucks lies not only in their power to redefine the identity of a classic American vehicle, but also in their contribution to a more sustainable and environmentally friendly future. As electric pickup trucks charge forward, they represent more than a change in power sources; they represent a break from tradition and a commitment to reshaping the automotive landscape for future generations. Electric pickup trucks are paving the way for a new era of sustainable haulage via innovation, competition, and a growing awareness of environmental responsibility.

The book aims to provide a comprehensive exploration of the electric pickup truck revolution, concentrating on Tesla's Cybertruck and its competitors such as Rivian's R1T, Ford F-150 Lightning, and GMC Hummer EV. The book's goal is to analyze the design, technology, performance, market effect, and future prospects of these electric giants, giving readers a thorough understanding of their importance in defining the automotive industry.

The Cybertruck: A Spectacle in Silicon Valley

Elon Musk and Tesla staged one of the most memorable events in automotive history—the unveiling of the Cybertruck—in the heart of Silicon Valley, where technological breakthroughs and daring ambitions converge. The world watched as Tesla's electric pickup truck, shrouded in secrecy and expectation, made its spectacular entrance on a crisp evening in November 2019. The event was more than simply a product reveal; it was a spectacle that transcended typical automotive unveilings, capturing the essence of Tesla's disruptive spirit and stretching the boundaries of design and expectation.

Prelude to the Unveiling: Excitement and Secrecy

In the days leading up to the event, excitement reached a fever pitch. Elon Musk, known for his enigmatic and sometimes unpredictable approach to product launches, teased the Cybertruck on social media, delivering glimpses that hinted at a departure from conventional truck design. The secrecy surrounding the vehicle fueled speculation and intrigue, attracting interest not just among automotive enthusiasts but also from the general public and media outlets throughout the world.

Tesla chose an appropriate location for the Cybertruck's grand debut—their enormous production site in Fremont,

California. The choice of the Fremont Factory, a location deeply woven into Tesla's history, added a layer of significance to the event. The area, where Tesla had overcome production challenges and evolved into a key player in the automotive sector, served as a symbolic background for the unveiling of a vehicle designed to redefine the concept of a pickup truck.

The fremont Factory was buzzing with excitement as the day of the unveiling approached. Tesla enthusiasts, industry insiders, and media representatives gathered in expectation of witnessing Tesla's next groundbreaking creation. The air was charged with electric excitement, matching the fervor that frequently accompanies the tech giant's major product launches.

Tesla's flamboyant and often contentious CEO, took center stage to unveil the Cybertruck. Musk, dressed in his signature casual style, exuded confidence as he addressed the audience. His opening remarks set the tone for the evening, emphasizing Tesla's commitment to pushing boundaries and challenging previously held beliefs about what a pickup truck might be. The gathering, a mix of Tesla supporters and skeptics, waited on every word, eager to see the unveiling of a vehicle shrouded in mystery.

The Cybertruck Emerges: A Daring Break from Tradition

As the moment of truth arrived, the Cybertruck rolled onto the stage, a vehicle unlike anything seen in the automotive industry before. Its angular, modern design defined classic pickup truck conveniences. The Cybertruck's exterior, crafted from cold-rolled stainless steel, gave it a striking and unconventional appearance, quickly separating it from its more conventional counterparts.

To emphasize the endurance and sturdy design of the Cybertruck, Elon Musk introduced a demonstration of the vehicle's "Armor Glass." The attempt to demonstrate the strength of the glass, however, took an unexpected turn when a metal ball hurled at the pane resulted in visibly visible cracks. Musk, ever the quick-witted, handled the issue with humour, turning the problem into a memorable moment that added an element of unpredictability to the event.

Aside from its unusual design, the Cybertruck's specifications were nothing short of outstanding. Musk unveiled a number of models, each with remarkable performance metrics. From acceleration to hauling capacity, the Cybertruck promised a level of power and versatility that pushed the boundaries of what a pickup truck could do. The electric drivetrain, a trademark of Tesla's vehicles, placed the Cybertruck in the forefront of the electric revolution in the pickup segment.

Design Philosophy

Elon Musk and his design team shared insights into the mindset that guided the Cybertruck's creation. The marriage of form and functionality became clear as they explained the aerodynamic considerations, the practicality of the exoskeleton, and the seamless incorporation of technology into the vehicle. Tesla's commitment to innovation and efficiency was reflected in the design choices.

The Cybertruck sprang from the thinking of Tesla's enigmatic CEO, Elon Musk, rather than standard design conventions. Musk, known for his daring and ambition, envisioned a vehicle that would not only defy expectations but also serve as a symbol of Tesla's commitment to breaking boundaries. The Cybertruck was created as a departure from the existing quo—a manifestation of Musk's vision for the future of both electric vehicles and automobile design.

The Cybertruck's design inspiration comes mostly from the realms of science fiction and futurism. Musk, a self-professed lover of cyberpunk aesthetics and futuristic narratives, wished to create a vehicle that appeared to have been driven directly out of a bleak future. The angular lines, metallic exoskeleton, and minimalist shape pay homage to classic vehicles from science fiction films, projecting an image of invention and otherworldliness.

The exoskeleton—an expanse of cold-rolled stainless steel—is at the core of the Cybertruck's design. This material, which is frequently used in industrial settings for its toughness and corrosion resistance, serves a dual purpose in the Cybertruck. It provides structural integrity similar to an armored vehicle while also contributing to the vehicle's aesthetic. Tesla's commitment to creating a sturdy, durable, and unique exterior is shown in the decision to use stainless steel.

The Cybertruck's angular geometry is the most eye-catching aspect of its design. Traditional pickup trucks have smooth curves and aerodynamic shapes, but the Cybertruck foregoes these conveniences in favor of jagged, polygonal lines. The careful selection of a geometric shape conveys a sense of edginess and modernity. It deviates from the organic shapes associated with vehicles, indicating Tesla's intent to redefine perceptions of what a pickup truck may be.

In contrast to the elaborate designs commonly seen in the automotive industry, the Cybertruck embraces simplicity. The absence of unnecessary embellishments and decorative elements emphasizes utility and purpose. Tesla's overarching concept of promoting efficiency and usability is consistent with the minimalist approach. Every line and shape has a specific purpose, adding to the overall operation of the vehicle.

The Cybertruck's design includes a continuous unibody structure, which eliminates the typical demarcation between the cabin and the cargo bed. This seamless integration adds to the vehicle's futuristic aesthetic while improving structural integrity. The unibody design, along with the stainless-steel exoskeleton, creates a unified and cohesive exterior that distinguishes the Cybertruck from typical pickup trucks.

Unlike other vehicles, the Cybertruck embraces its raw materiality rather than elaborating paint processes to achieve a glossy look. During the unveiling, Elon Musk emphasized the vehicle's resistance to dents and scratches, emphasizing the practicality of its stainless-steel exterior. The decision to forego standard color options is consistent with the Cybertruck's utilitarian ethos, presenting a vehicle that prioritizes substance over superficial aesthetics.

The Cybertruck wasn't just about looks; it also had several unusual features that set it different from other pickup trucks. The adaptive air suspension, capable of modifying the vehicle's ride height on the fly, demonstrated Tesla's commitment to combining performance with flexibility. The vault-like storage compartment, located at the rear of the vehicle, added a practical and secure element to the design of the Cybertruck.

Finally, the Cybertruck's design is more than just a jumble of stainless steel and angular lines; it represents a revolution in automotive design philosophy. Tesla's bold departure from established conventions, exemplified by the Cybertruck, challenges the industry to reimagine the possibilities of vehicle aesthetics. The controversies and debates sparked by the Cybertruck's design are not isolated incidents, but rather reflections of a broader debate about the future of transportation and the role of design in creating that future. Whether celebrated as a symbol of innovation or chastised for deviating from convention, the Cybertruck is a testament to Tesla's commitment to pushing boundaries, generating dialogue, and ushering in a new era in automotive design.

Reactions and Controversies

The unveiling of the Tesla Cybertruck was more than just an automotive event; it was a cultural moment that sparked a wave of reactions and controversies. The Cybertruck rapidly became a matter of intense debate and discussion, from its unconventional design to the unexpected "Armor Glass" demonstration. This chapter delves into the first reactions and controversy surrounding the Cybertruck, examining the public's reaction, media scrutiny, and Tesla's unexpected challenges in the aftermath of the big reveal.

The Unveiling: Surprise and Awe

As the Cybertruck rolled onto the stage at Tesla's Fremont Factory, the audience let out a collective gasp. The angular, stainless-steel exoskeleton, resembling a vehicle from a sci-fi dystopia, contrasted sharply with the classic curves and shapes of conventional pickup trucks. The audience's reaction, which included Tesla enthusiasts, journalists, and industry insiders, ranged from horror to awe. The Cybertruck was more than simply a car; it was a visual and conceptual departure from everything the automotive world had come to expect.

Memes and Mockery

Within minutes of the Cybertruck's unveiling, social media platforms erupted with memes, jokes, and a frenzy of commentary. The unusual design, which some compared to a polygonal rendering or a vehicle from a retro-futuristic video game, became a goldmine for internet fun. Images and analogies flooded the internet, with references to movies, cartoons, and even objects that somewhat resembled the Cybertruck's distinct silhouette. The Cybertruck's power to capture public attention and encourage creativity, even if in the form of satire, was demonstrated by the instantaneous and widespread reaction on social media.

The "Armor Glass" Incidence.

The "Armor Glass" demonstration was one of the Cybertruck unveiling's defining moments. Elon Musk attempted to demonstrate the robustness of the vehicle's glass by throwing a metal ball at them. However, to everyone's surprise, the ostensibly shatterproof glass cracked on contact. Musk handled the issue with humour, but the incident became a focus point for detractors and skeptics. The unexpected incident underscored the difficulties of live demonstrations and the unpredictability of new unveilings.

Aesthetic Controversies

The design of the Cybertruck was a significant departure from the established rules of pickup truck aesthetics. While some admired its futuristic and edgy appearance, others considered it divisive and even unappealing. Traditional pickup trucks, defined by curves, chrome accents, and a rugged yet familiar silhouette, had defined the segment for decades. The Cybertruck, with its geometric lines and sharp angles, challenged these conventions, sparking debates about the balance between innovation and adherence to established design expectations.

Functionality vs. Form

In addition to the controversy surrounding its appearance, the Cybertruck's design was scrutinized for its practicality and usefulness. The stainless-steel exoskeleton, described as "armor" by Elon Musk, raised concerns about the vehicle's weight, manufacturing costs, and repairablilty. Critics contended that the design prioritized form above function, calling Tesla's claim that the exoskeleton offered both durability and utility into question. In the aftermath of the unveiling, the clash between aesthetics and pragmatism became a central issue of discussion.

In the days and weeks after the Cybertruck's unveiling, industry experts and observers shared their thoughts on the event. Successes such as the attention-grabbing design, the strong specifications, and the impressive preorder numbers were discussed alongside challenges such as the unexpected glass demonstration hiccup and concerns regarding market reception. The Cybertruck has successfully sparked discussions on the future of pickup trucks and the role of electric vehicles in influencing the automotive landscape.

As the dust settled from the unveiling spectacle, the automotive industry shifted its focus to the Cybertruck's future prospects and its potential influence on the pickup segment. Will a broader audience embrace the unconventional design? How will competitors react to Tesla's entry into the electric truck market? As the

Cybertruck set the stage for a new chapter in the continuous evolution of pickup trucks, these questions lingered.

The unveiling of the Cybertruck in Silicon Valley was a watershed moment in automotive history. Aside from the shock and awe of its unconventional design, the event encapsulated Tesla's ethos of challenging norms, pushing boundaries, and driving innovation. The Cybertruck became more than simply a vehicle; it became a symbol of the automotive industry's ever-evolving nature and the daring required to reshape the future. As the Cybertruck entered the spotlight, it left an indelible mark, igniting conversations that reverberated far beyond the confines of the Fremont Factory and Silicon Valley—a spectacle that will be remembered as a milestone in the journey toward a electrified automotive future.

Elon Musk, known for his active presence on social media, did not shy away from addressing the Cybertruck controversy. Instead, he participated in the debates, responding to criticisms, giving behind-the-scenes details, and keeping an open dialogue with the public. Musk's approach, a blend of transparency, humor, and unwavering confidence in the Cybertruck's mission, added a layer of authenticity to Tesla's engagement with its audience. The controversy surrounding the Cybertruck did not diminish its exposure; rather, it was propelled into the forefront. The "Streisand effect," a phenomenon in which attempts to conceal, remove, or control information result in more notoriety, came into play. The more the Cybertruck became the subject of debate and discussion, the more it captured the public's attention. Tesla's bold venture drew even more eyes.

Beyond Convenience: Tesla's Vision for the Future of Pickup Trucks

While the early reactions and debates around the Cybertruck were dominated by its design, Tesla's vision for the future of pickup trucks goes well beyond aesthetics. This section delves into Tesla's overarching vision, examining the technological advances, environmental aims, and market strategies that underpin the Cybertruck's role in reshaping the pickup truck landscape.

Technological Abilities.

The commitment to improving electric vehicle technology lies at the heart of Tesla's ambition for the Cybertruck. The Cybertruck has an all-electric drivetrain, putting an end to the reliance on internal combustion engines that has defined the pickup truck segment for decades. Tesla's expertise in electric powertrains, as seen by the success of their previous models, places the Cybertruck as a technological flagship, marking a transition toward cleaner, more sustainable transportation.

Performance Metrics

Aside from its eye-catching design, the Cybertruck's performance metrics are a testament to Tesla's aim for redefining pickup truck capabilities. The vehicle's acceleration, towing capacity, and off-road capabilities demonstrate the might of electric power in exceeding traditional expectations. Tesla's goal is to not only compete in the pickup market, but to outperform it, establishing a new bar for what pickup trucks can achieve in terms of power and efficiency.

Sustainable Materials

Tesla's commitment to sustainability extends to the materials used in the building of the Cybertruck. The stainless-steel exoskeleton is chosen for its recyclability as well as its durability. Tesla's dedication to reducing environmental effect goes beyond the exterior; the interior of the Cybertruck features sustainable materials, contributing to the company's broader objective of creating vehicles with a smaller ecological footprint.

Autonomy and Driver-Assist Features.
The Cybertruck is equipped with advanced driver-assist features and the potential for complete self-driving capabilities, as part of Tesla's vision for the future of pickup trucks. The incorporation of Tesla's advanced driver-assistence system Autopilot is a step toward autonomous driving. While

regulatory hurdles and technological challenges remain, Tesla envisions a future in which the Cybertruck, and vehicles in general, are more than modes of transportation—they are intelligent, connected entities that enhance the overall driving experience.

Shift in Market Dynamics

Tesla's ambition for the Cybertruck goes beyond appealing to conventional pickup truck enthusiasts. Tesla aspires to appeal to a diverse audience by mixing performance, sustainability, and cutting-edge technology. While originally divisive, the Cybertruck's unconventional design presents it as a symbol of innovation and a departure from the established quo. Tesla envisions a market in which the Cybertruck not only competes with other pickup vehicles, but also draws consumers who have never considered a pickup truck before.

Cybertruck's Role in Tesla's Product Lineup

The Cybertruck plays a strategic role in diversifying Tesla's offerings in the broader context of the company's product lineup. While Tesla rose to prominence with its electric sedans, the unveiling of the Cybertruck signifies the company's entry into the lucrative pickup truck market. Tesla envisions a future in which the Cybertruck contributes considerably to its overall market share, cementing the company's dominance in the electric vehicle space.

The Influence on the Automotive Landscape

Tesla's ambition for pickup truck success beyond the success of a single vehicle. The Cybertruck's daring design and innovative features act as a catalyst for change in the automotive landscape. Competitors are taking notice, and the industry is being forced to reconsider old approaches to design, technology, and market strategy. Tesla envisions a future in which the Cybertruck's effect extends beyond its sales figures, influencing the pickup truck segment as a whole.

Responding to Skepticism and Evolving Perceptions

Tesla remains steadfast in its vision for the Cybertruck in the face of skepticism and criticism. The business recognizes that groundbreaking designs frequently face resistance at first, but believes that the Cybertruck's unique features and capabilities will triumph over consumers in the long run. Tesla envisions a paradigm shift in which the Cybertruck becomes more than simply an outlier, but a trailblazer who raises the bar for what a pickup truck can do.

Tesla's idea for the Cybertruck is fraught with difficulties. The development of a vehicle with a stainless-steel exoskeleton and an unconventional design poses unique production challenges. Navigating regulatory landscapes, particularly in terms of autonomous driving capabilities, necessitates partnership with governing bodies. Tesla also recognizes the need to address consumer concerns and foster widespread acceptance of electric pickup trucks. The vision entails overcoming these obstacles, setting the way for a future in which the Cybertruck becomes a symbol of success and innovation.

Tesla's vision for the future of pickup trucks goes beyond the Cybertruck. It heralds a broader shift in transportation—an era in which electric vehicles, autonomy, and sustainability redefine how we move. Tesla envisions a future in which the Cybertruck's success contributes to a larger shift in consumer preferences, industry conventions, and the overall trajectory of the automotive landscape.

In summary, the first reactions and controversy surrounding the Cybertruck were not just reactions to an unconventional vehicle design; they were reactions to Tesla's visionary leap into the future of pickup trucks. The controversies, debates, and unexpected challenges become integral aspects of a larger narrative—a narrative that extends beyond a single vehicle and encapsulates Tesla's commitment to breaking boundaries, reshaping industries, and driving innovation. As the Cybertruck generates debate and inspires change, it serves as a symbol of Tesla's vision for a future in which pickup trucks are more than just modes of transportation, but pioneers of a transformative era in automotive history.

Electric Pickup Models: Transforming the Trucking Landscape

The automotive sector, formerly synonymous with roaring engines and gas-guzzling behemoths, is undergoing a significant shift as a result of the emergence of electric vehicles (EVs). Electric pickup trucks are taking center stage in this transformation, challenging the traditional concept of rugged, gas-powered pickups. In this exploration, we look at three notable electric pickup models: the Rivian R1T, the Ford F-150 Lightning, and the GMC Hummer EV. With their own features, design philosophies, and market strategies, these electric titans are reshaping the shipping landscape and paving the way for a more sustainable future.

Rivian R1T: The Daring Trailblazer

The Rivian R1T is a testament to the potential of electric pickups in terms of performance, advancement, and sustainability. Rivian, a relatively new American automaker, has quickly risen to prominence with its concentration on electric adventure vehicles. Rivian's flagship model, the R1T, embodies the company's commitment to pushing the boundaries of electric mobility.

Design and Creativity

The R1T's design concept reflects a harmonic blend of rugged capacity and modern aesthetics. The R1T maintains a sense of refinement while retaining the powerful characteristics expected of a pickup truck, thanks to a sleek front fascia and striking lighting elements. The lack of a standard internal combustion engine allows for creative design choices, such as the "gear tunnel" that runs through the body of the truck, giving extra storage for outdoor enthusiasts.

Performance Metrics

Rivian markets the R1T as an off-road powerhouse with exceptional performance metrics. The quad-motor setup, one for each wheel, not only contributes to exceptional off-road capabilities, but also ensures precise control and increased stability. The R1T's acceleration figures are notable, challenging the stereotype that electric vehicles sacrifice power.

Adventure ready features

What distinguishes the Rivian R1T is its emphasis on adventure. The truck has an air suspension system that may raise or lower the vehicle depending on the terrain. Rivian also provides a comprehensive Off-Road Upgrade, which includes features such as underbody protection, off-road tires, and a reinforced undercarriage. These features turn the R1T become more than just an electric pickup; they invite you to explore the great outdoors.

Range of Batteries and Charging Infrastructure

The R1T addresses the critical aspect of range anxiety with an excellent battery range. Rivian provides many battery options, allowing customers to choose between different ranges based on their needs. Rivian is also investing in its charging infrastructure, with ambitions to establish the Rivian Adventure Network—an expansive network of charging stations strategically placed near outdoor destinations.

Positioning on the Market

Rivian's marketing strategy is around promoting the R1T as a high-end adventure vehicle. The R1T aspires to tap into a niche market that values both sustainability and off-road capabilities, with a focus on outdoor enthusiasts. Rivian's approach emphasizes not only the vehicle but also the way of life it represents—a blend of luxury, adventure, and environmental responsibility.

Ford F-150 Lightning: A Legend goes electric

When one thinks about pickup trucks, the Ford F-150 is unavoidable. The F-150 has been a symbol of American automotive prowess for decades, remaining the best-selling vehicle in the United States. With the debut of the F-150 Lightning, Ford enters the electric era, blending the legacy of the F-150 with the innovation of electric power.

Electric Power Integration Seamless

The F-150 Lightning keeps the traditional design language of the F-150 series while seamlessly integrating electric power. From the outside, the Lightning is indistinguishable from its gas-powered counterparts, preserving the rugged aesthetic that F-150 fans like.

However, the electric powertrain opens up new possibilities in terms of performance and usefulness.

Towing Capacity and Performance

The F-150 Lightning's performance capabilities have not been compromised by Ford. The electric powertrain converts to immediate torque, offering tremendous acceleration. Furthermore, the F-150 Lightning has exceptional towing capabilities, putting to rest the misconception that electric trucks fall behind their gas-powered counterparts in terms of usefulness. This emphasis on performance ensures that the Lightning meets the expectations set by its F-150 predecessors.

Onboard Power and Versatility

The Pro Power Onboard system, which essentially converts the truck into a mobile power generator, is one notable feature of the F-150 Lightning. The Lightning, with its many power output possibilities, may serve as a source of electricity for tools, appliances, and even as a emergency power supply. This level of versatility increases the truck's appeal, especially for those who rely on their vehicles for business and recreation.

Technology and Linked Features

The F-150 Lightning features modern connectivity and technology. The Lightning, with its large touch-screen infotainment system and advanced driver-assist technologies, ushers the F-150 into the digital age. The addition of these features by Ford not only improves the driving experience, but also puts the F-150 Lightning as a technologically advanced alternative in the electric pickup segment.

Positioning and Accessibility on the Market

The F-150 Lightning's market strategy is around accessibility and widespread acceptance. Ford hopes to attract a broad audience by retaining the classic F-150 design and embracing electric power, including existing F-150 enthusiasts who may be considering the change to electric vehicles. The F-150 Lightning is a watershed moment in Ford's history, a significant step toward electrifying its iconic truck.

GMC Hummer EV: Bringing Back an Icon in Electric Form

The Hummer, once a emblem of military-inspired ruggedness, is making a surprise return in the shape of the GMC Hummer EV. While the original Hummer was well-known for its gas-guzzling nature, the new iteration embraces electric power, indicating a paradigm change for this famous brand.

Size and Design

The GMC Hummer EV maintains the dominating presence and strong design that has been synonymous with the Hummer name. It is, in essence, a supertruck, distinguished by its imposing size and peculiar aesthetics. The design incorporates modern elements that reflect its electric identity while nodding to classic Hummer flair. The Hummer EV, which is available in

SUV and truck versions, caters to a wide range of consumer preferences.

Performance both on and off the road

The Hummer EV is positioned by GMC as a electric supertruck with unrivaled performance capabilities. The Hummer EV has a three-motor setup that provides exceptional power and torque distribution to each wheel. This arrangement not only produces excellent on-road performance, but it also contributes to the Hummer EV's off-road prowess. GMC emphasizes the Hummer EV's capability in difficult terrain, making it a compelling alternative for adventure seekers.

CrabWalk and Extract modes

The CrabWalk Mode and Extract Mode are two of the Hummer EV's notable features. The CrabWalk Mode allows the vehicle to move diagonally, improving maneuverability in small spaces. On the other hand, Extract Mode raises the suspension by several inches, allowing the Hummer EV to navigate obstacles with ease. These features demonstrate GMC's commitment to innovation and usefulness, distinguishing the Hummer EV in terms of both form and capabilities.

Ultium Battery Technology

The Hummer EV's performance is powered by General Motors' Ultium battery technology. This advanced battery system not only provides the required power for the Hummer EV's electric motors, but also contributes to impressive range figures. The Ultium technology is part of General Motors' larger strategy to accelerate the adoption of electric vehicles across its brands.

Interior Design and Entertainment

GMC did not skimp on the inside of the Hummer EV, injecting it with opulent elements and cutting-edge technology. Premium materials, advanced infotainment choices, and a driver-centric layout are included in the cabin. The mix of comfort and technology improves the whole driving experience, establishing the Hummer EV as a high-end alternative in the electric pickup market.

Positioning in the Market as an Electric Supertruck

The Hummer EV's market positioning by GMC revolves around its identity as a electric supertruck. GMC hopes to attract consumers looking for a premium, high-performance electric pickup by elevating the legendary Hummer brand and reimagining it as a electric powerhouse. The Hummer EV is more than a vehicle; it's a work of art—a symbol of the convergence of elegance, power, and sustainability.

Comparative Analysis: Common Paths and Distinct Selling Points

Technology of Range and Battery

One of the most important aspects influencing the success of electric vehicles is the range provided by their batteries. In this regard, all three models—the Rivian R1T, the Ford F-150 Lightning, and the GMC Hummer EV—use advanced battery technology to deliver impressive range figures. Rivian provides many battery options for the R1T, allowing customers to select the range that best meets their needs.

Ford's F-150 Lightning focuses on versatility while retaining competitive range figures by leveraging its Pro Power Onboard system. The GMC Hummer EV, which comes standard with GM's Ultium battery technology, stands out for its emphasis on both performance and range.

Towing Capacity and Performance

Considering the historic link of pickup trucks with power and usefulness, performance is an important aspect of electric pickup vehicles. The Rivian R1T's quad-motor setup provides exceptional off-road capability and acceleration. Ford's F-150 Lightning, true to the F-150 series' legacy, has exceptional towing capability and quick torque from its electric powertrain. The GMC Hummer EV, positioned as a supertruck, excels in both on-road and off-road performance, demonstrating the potential of electric power in the realm of high-performance pickups.

Innovative features and technological integration

Each model brings a distinct set of innovative features to the table. Rivian's R1T is designed for adventure, featuring features like as a gear tunnel and off-road upgrades. The Pro Power Onboard system is introduced in Ford's F-150 Lightning, converting the truck into a mobile power generator. GMC's Hummer EV distinguishes itself with CrabWalk Mode and Extract Mode, demonstrating the brand's commitment to innovation in both utility and design. All three models have advanced technology and connectivity features, reflecting the industry's trend toward a more digital and connected driving experience.

Target Audience and Market Strategies

The market strategies used by Rivian, Ford, and GMC differ, reflecting different approaches to the electric pickup segment. Rivian offers the R1T as a premium adventure vehicle, aimed at outdoor enthusiasts who value sustainability as well as off-road capabilities. Ford's F-150 Lightning prioritizes accessibility and mass adoption, using the legendary F-150 brand to appeal to a diverse audience, including current F-150 owners. GMC's Hummer EV takes a premium approach, portraying itself as a electric supertruck with a blend of luxury, power, and sustainability, appealing to consumers looking for a high-performance electric pickup.

The Electric Pickup Landscape: Implications and Future Trends

The emergence of electric pickup trucks represents a big change in the automotive landscape. Beyond the individual success of models such as the Rivian R1T, Ford F-150 Lightning, and GMC Hummer EV, the collective influence is reshaping perceptions and propelling the industry toward a more sustainable future.

Transportation Sustainability and Environmental Impact

One of the key motivators driving the development of electric vehicles, particularly pickups, is the desire to reduce the environmental effect of transportation. The transition from internal combustion engines to electric power reduces carbon emissions dramatically, adding to worldwide efforts to mitigate climate change. As electric pickup trucks acquire market momentum, their widespread adoption might result in a significant reduction in the carbon footprint associated with the trucking business.

Charge Infrastructure Challenges and Opportunities

The success of electric vehicles is heavily dependent on the availability and accessibility of charging infrastructure. Automakers are investing in charging networks to alleviate range anxiety and encourage the use of electric pickups, as shown in Rivian, Ford, and GMC strategies. The development of a strong charging infrastructure is not only a challenge, but also an opportunity for collaboration between automakers, energy companies, and governments to create a seamless and wide-spread charging network.

Consumer Adoption and Changing Preferences

The success of electric pickup trucks is dependent on consumer acceptance, which is influenced by shifting perceptions. While enthusiasts and early adopters may embrace the novelty and performance of electric pickups, broader consumer acceptance necessitates overcoming skepticism and dispelling misunderstandings about electric vehicles. The success stories of models such as the Rivian R1T, Ford F-150 Lightning, and GMC Hummer EV help to reshape perceptions and foster a more positive attitude toward electric trucks.

Competitive Dynamics and Industry Reaction

The entry of well-known automakers like as Ford into the electric pickup market enhances competition and accelerates innovation. As electric pickups become more popular, conventional automakers are being compelled to develop their electric offerings in order to remain competitive. This dynamic is fueling a broader industry-wide trend toward electrification, affecting not only pickups but also other vehicle segments.

Landscaping Regulations and Government Incentives

Government policies and incentives play an important role in influencing the electric vehicle market. Regulations aimed at lowering emissions and encouraging environmentally friendly transportation contribute to the emergence of electric pickups. Incentives like as tax credits and rebates may have a big influence on consumer decisions, making electric pickups more financially appealing. As governments throughout the world prioritize environmental sustainability, the regulatory landscape is likely to encourage the use of electric vehicles even more.

In conclusion, The Rivian R1T, Ford F-150 Lightning, and GMC Hummer EV are not simply individual vehicles, but also forerunners in the electric pickup era.

Each model brings its own unique strengths, design philosophies, and market strategies to the table, together creating trucking's future. These electric pickup trucks serve as emblems of innovation and sustainability as they navigate challenges, break new ground in technology, and redefine consumer expectations. The electric pickup landscape, formerly a niche segment, is now at the forefront of automotive evolution, setting the way for a future in which the roar of engines is replaced by the hum of electric motors and the trucking sector enters a new era of electrification.

The Electric Titans: Powertrains, Performance Metrics, and Range Comparison

The rise of electric vehicles (EVs) is causing a seismic upheaval in the automotive landscape, notably in the pickup truck segment. As Tesla's Cybertruck prepares to enter the market, it will face formidable competition from other electric titans, each with unique powertrains, performance metrics, and supreme range capabilities. In this in-depth exploration, we dissect the powertrains, delve into the performance metrics, and compare the range of the Tesla Cybertruck to its counterparts—Rivian R1T, Ford F-150 Lightning, and GMC Hummer EV.

Powertrains: Elevating the Pickup Truck Segment

Tesla Cybertruck: Tri-Motor Powerhouse
With its advanced powertrain choices, the Tesla Cybertruck is poised to revolutionize the pickup truck segment. The Cybertruck's flagship version has a tri-motor setup, with independent motors for the front and rear axles and an extra motor for the rear axle. This design helps the Cybertruck's exceptional acceleration, off-road capability, and overall performance. The efficiency and power delivered by the tri-motor setup exemplify Tesla's expertise in electric powertrains.

Rivian R1T: Quad-Motor Superiority

Rivian takes a unique approach with the R1T, which has a quad-motor setup—one motor for each wheel. This design not only improves the R1T's off-road capabilities, but also provides precise control over each wheel, enhancing performance in a variety of driving circumstances. The quad-motor setup adds to the R1T's reputation as an adventure-ready electric pickup capable of conquering difficult terrain with ease.

Dual-Motor Innovation in the Ford F-150 Lightning

The F-150 Lightning, Ford's entry into the electric truck market, has a dual-motor setup. While the precise details of the powertrain combinations have yet to be revealed, the dual-motor setup is expected to deliver a balance of performance and efficiency. Ford hopes to seamlessly integrate electric power into the F-150 Lightning without sacrificing its legendary capabilities, using its experience in producing conventional pickup trucks.

GMC Hummer EV: THREE-Motor Supremacy

The GMC Hummer EV has a three-motor powertrain setup that combines power and precision for on-road and off-road performance. The Hummer EV, like the Cybertruck, has independent motors for the front and rear axles, as well as an extra motor for the rear axle. This layout, in combination with advanced torque distribution capabilities, contributes to the Hummer EV's exceptional acceleration and maneuverability.

Performance Metrics: Acceleration, Towing, and Off-Road Capability

Tesla Cybertruck: Absurd Acceleration

Tesla is well-known for its Ludicrous Mode, a feature that pushes the limits of electric vehicle acceleration. The Cybertruck, outfitted with a tri-motor setup, is expected to demonstrate Ludicrous Mode, delivering amazing acceleration figures. The exact metrics for the Cybertruck's acceleration have not been officially disclosed, but Tesla's track record suggests that it will be a formidable force on the road, combining speed with the instantaneous torque characteristic of electric powertrains.

Rivian R1T: Off-Road Capability

Rivian markets the R1T as an adventure vehicle, and its performance metrics reflect this. The quad-motor setup contributes to the R1T's off-road prowess by providing for precise control and torque distribution to each wheel. Rivian highlights not just acceleration but also the R1T's ability to navigate difficult off-road terrain. The R1T's performance metrics are designed to appeal to outdoor enthusiasts looking for a versatile and capable electric pickup.

Workhorse Capability: Ford F-150 Lightning

The F-150 Lightning, according to the F-150 series' legacy, is expected to provide strong towing capabilities. While specific towing metrics have yet to be revealed, Ford intends to promote the F-150 Lightning as a workhorse capable of meeting the demands of typical F-150 owners. The dual-motor setup is likely to strike a balance between performance and utility, making the F-150 Lightning a versatile option for work and pleasure.

GMC Hummer EV: Off-Road Precision
The GMC Hummer EV's three-motor setup not only helps with acceleration but also improves its off-road capabilities. GMC has emphasized features such as CrabWalk Mode, which allows the vehicle to move diagonally, demonstrating the precision and agility of the Hummer EV in off-road scenarios. While specific acceleration figures have yet to be released, the Hummer EV's performance metrics are consistent with its identity as a supertruck designed for both on-road and off-road adventures.

Comparison of Ranges: The Need for Long-Distance Travel

Tesla Cybertruck: Benchmark Range

Tesla has set the standard for electric vehicle range, and the Cybertruck is expected to follow suit. The range of the Cybertruck varies depending on the battery combination used. Tesla offers three configurations: Single Motor RWD, Dual Motor AWD, and Tri-Motor AWD, with estimated ranges exceeding 500 miles for the Tri-Motor AWD. Tesla's expertise in battery technology and energy efficiency resulted in the Cybertruck's impressive range.

Rivian R1T: Varied Range Options

Rivian allows buyers to choose from many battery combinations for the R1T, allowing them to adjust the range to their own needs. The R1T is available in three battery sizes—105 kWh, 135 kWh, and 180 kWh—to cater to a wide range of users. The largest battery option, at 180 kWh, is expected to provide an excellent range, placing the R1T as a versatile alternative for long-distance travel.

Ford F-150 Lightning: Striking a Balance Between Range and Utility

Ford's approach to the F-150 Lightning's range is based on balancing the need for a larger range with the truck's usability and performance. While specific range figures have yet to be released, Ford intends to offer a competitive range that meets F-150 owners' expectations. The range of the F-150 Lightning will most likely reflect a strategic compromise between long-distance travel capabilities and the demands of a regular pickup truck.

GMC Hummer EV: Surprising Range

The GMC Hummer EV, positioned as a supertruck, is expected to deliver an impressive range that corresponds to its high-performance capabilities. GMC has not revealed specific range figures for the Hummer EV, but it is expected to have a competitive range given its three-motor setup and advanced battery technology. The Hummer EV's range is likely to attract drivers looking for both power and the ability to cover long distances on a single charge.

Charging Infrastructure: Electric Mobility's Backbone

Tesla Supercharging Network: Unrivaled Charging Speeds

Tesla's proprietary Supercharger network is one of the company's key advantages in the electric vehicle market. Tesla Superchargers are noted for their high charging speeds, allowing Tesla vehicles to replenish a substantial chunk of their battery capacity in a short period of time. As a Tesla vehicle, the Cybertruck benefits from access to this expansive and fast-charging network, increasing its utility for long-distance travel.

Rivian Venture Network: Extending Charging Capability

Rivian intends to address its R1T customers' charging needs via the Rivian Adventure Network. This network of charging stations is strategically placed around outdoor destinations, catering to Rivian owners' adventurous mentality. While the charging speeds may not be as fast as Tesla's Superchargers, Rivian's emphasis on charging accessibility adds to the R1T's appeal for those seeking off-road adventures and long-distance travel.

FordPass Charging Network: National Convenience

Ford leverages its expansive FordPass charge Network to provide F-150 Lightning owners with easy access to charge stations around the country. While charging speeds may vary depending on infrastructure, Ford's commitment to establishing a comprehensive charging network matches with the company's objective of making electric mobility accessible to a wide range of people. The charging infrastructure of the F-150 Lightning contributes to its attractiveness as an electric workhorse.

GMC's Charging Ecosystem Ultium Charge 360

GMC, a division of General Motors, introduces Ultium Charge 360, a ecosystem aimed at simplifying the charging experience for Hummer EV owners. This ecosystem includes access to numerous charging providers, public charging stations, and the Ultium Charge 360 app for a seamless charging experience. While details concerning charging speeds have yet to be revealed, GMC's comprehensive approach to charging infrastructure increases the Hummer EV's utility for electric travel.

Comparative Analysis: Weaknesses and Trade-Offs

Tesla Cybertruck: Benchmark Range and Charging Speeds

The Tesla Cybertruck distinguishes out with its benchmark range, offering versions that exceed 500 miles. The Cybertruck's integration with Tesla's Supercharger network ensures speedy charging speeds, making it a compelling alternative for those who value long-distance travel and quick charging stops. Tesla's expertise in battery technology and charging infrastructure gives the Cybertruck a strategic advantage in the electric pickup market.

Rivian R1T: Flexible and Adventure-Ready Charging

Rivian's R1T excels in versatility, with many battery combinations to cater to various range requirements. While not matching Tesla's charging speeds, the Rivian Adventure Network adds to the R1T's appeal for adventure enthusiasts. Rivian offers the R1T as a electric pickup that combines range options with off-road capabilities, giving it a versatile solution for a wide range of users.

Ford F-150 Lightning: Wide Popularity and Nationwide Charging

The F-150 Lightning's strength is its broad appeal, which caters to a wide range of F-150 owners with diverse needs. The F-150 Lightning's convenience for long-distance travel is enhanced by Ford's expansive FordPass Charging Network. While the charging speeds may not be as fast as Tesla's Superchargers, Ford's aim on establishing a comprehensive charging network reflects the company's commitment to making electric transportation accessible to a wider audience.

GMC Hummer EV: Surprising Performance and Charging Ecosystem

The GMC Hummer EV advertises itself as a supertruck, and its range is expected to reflect its high-performance capabilities. Although specific charging speeds are yet to be detailed, GMC's Ultium Charge 360 ecosystem provides a comprehensive charging solution. The Hummer EV caters to drivers looking for power as well as a well-integrated charging experience, making it a premium choice in the electric pickup market.

Final Thoughts

Finally, the electric pickup segment is witnessing a revolution, with each contender—Tesla Cybertruck, Rivian R1T, Ford F-150 Lightning, and GMC Hummer EV—bringng its strengths to the table. The powertrains, performance metrics, and range capabilities of these electric titans represent a fundamental shift in the automotive industry. Tesla's Cybertruck, benchmarked by its Supercharger network, sets a high standard for long-distance travel.

Rivian's R1T emphasizes versatility and adventure-ready capabilities, aimed for a wide range of users. Ford's F-150 Lightning has widespread appeal, thanks to a nationwide charging network, and is aimed at regular F-150 owners. GMC's Hummer EV, positioned as a supertruck, focuses on excellent performance and a well-integrated charging ecosystem.

As these electric giants prepare to reach the road, they not only represent a change toward sustainability, but they also challenge previously held beliefs about pickup trucks' capabilities and desirability. The next years will see more advancements, innovations, and the further evolution of electric powertrains, stretching the boundaries of what is possible in the pickup truck segment. Whether it's the acceleration of the Cybertruck, the off-road prowess of the R1T, the workhorse capabilities of the F-150 Lightning, or the supertruck performance of the Hummer EV, these electric titans collectively shape the future of pickup trucks, electrifying the landscape and paving the way for a new era in automotive mobility.

Personal reflection

Hummer EV

The concept of 'safety' remains subjective, with my concerns revolving not just around personal safety, but the safety of fellow road users when it comes to electric vehicles. These vehicles, owing to their hefty batteries, generally weigh more than their gas-powered counterparts of similar size, a characteristic glaringly exemplified by the extreme case of the Hummer EV Edition 1.

While General Motors has indicated a production of over 65, 000 Hummer EVs, most won't mirror the Edition 1's massive specifications. However, the likelihood of them being considerably heavy raises concerns, particularly regarding visibility issues. The correlation between weight and safety in car crashes is rooted in basic physics; the heavier object tends to dominate in collisions,   posing higher risks to occupants of lighter vehicles.

Even in scenarios where the weight difference is relatively marginal,   the disadvantages for occupants in the lighter vehicle are evident. Insurance claim data supports this, emphasizing the safety trade-off where those in heavier vehicles are less likely to be injured at the expense of those in lighter vehicles. The extent to which the danger to other vehicles escalates with an increasing weight gap, particularly in the context of behemoths like the Hummer EV, remains uncertain and requires real-world crash data for quantification.

The responsibility for safety doesn't solely rest on the weight of the vehicle but also extends to the behavior of the drivers. High-horsepower vehicles, irrespective of their weight, are statistically more prone to serious crashes, raising questions about the driving habits of Hummer owners. The potential conscientiousness of these drivers may, to some extent, offset safety concerns associated with the vehicle's weight.

General Motors, cognizant of the potential challenges posed by the Hummer Edition 1's weight, asserts that its engineers diligently addressed this issue during the development process. The incorporation of standard safety technologies, including pedestrian detection and lane-keeping assistance, aims to mitigate risks. Additionally, features like Super Cruise, touted as one of the safest hands-free highway driving assistance systems, further contribute to the safety net during drives.

However, it's essential to scrutinize specific aspects of safety. For instance, the Hummer Edition 1's stopping distance, claimed to be 157 feet from 60 miles per hour, while impressive, falls short of the average pickup truck's stopping distance of 140 feet. This discrepancy raises questions about the Hummer's ability to handle emergency braking situations efficiently, especially given its weight.

The issue of weight extends beyond the Hummer EV Edition 1. Other electric trucks, such as the Ford F-150 Lightning and Rivian R1T, also boast substantial weights. The Ford F-150 Lightning, even without passengers or cargo, tips the scales at about 6, 500 pounds, while the Rivian R1T exceeds 7, 000 pounds. These weights, when compared to traditional vehicles like the Chevrolet Tahoe or a compact car like the Nissan Sentra, underscore the considerable heft of electric trucks.

The challenge posed by extreme weight extends to regulatory frameworks. Vehicle safety regulations in the United States were primarily designed around the assumption that exceptionally heavy vehicles would be commercial trucks rather than everyday vehicles. The Hummer EV's Gross Vehicle Weight Rating (GVWR) of 10, 550 pounds subjects it to different safety rules, impacting features like side mirrors. Federal regulations mandate flat mirrors instead of curved ones for vehicles with a GVWR exceeding 10, 000 pounds, affecting visibility and potentially creating blind spots.

During a test drive of the Hummer EV, the flat side mirror raised concerns about significant blind spots, necessitating heavy reliance on the Blind Spot Monitor. Recommendations for improvements, such as making the Blind Spot Monitor light more prominent, aim to address safety concerns arising from altered mirror configurations mandated by weight-related regulations.

In conclusion, the safety implications of the weight of electric trucks, exemplified by the Hummer EV Edition 1, extend beyond the vehicle's performance on the road. The intricate interplay between weight, safety features, and regulatory requirements necessitates a nuanced evaluation. As electric trucks become a prominent feature on roadways, it becomes imperative to address safety concerns comprehensively, incorporating not only vehicle design and engineering but also driver behavior and adherence to safety regulations. This multifaceted approach is essential for fostering the safe integration of electric trucks into the broader automotive landscape.

Ford EV

Swift, powerful, and exceeding the performance of any previous F-150, the Lightning stands as a testament to innovation, offering functionalities that redefine the conventional notion of a truck. Beyond the usual capacities, this electric marvel transforms into a mobile workspace with an unexpected twist—a drinks cooler under its watertight front storage compartment, complete with drain plugs. Moreover, it possesses the capability to power an entire household for an extended period, showcasing versatility beyond the traditional truck's scope.

Contrary to being a mere camping accessory or an eccentric design experiment, the F-150 Lightning is a genuine workhorse. Currently in production by a company well-versed in large-scale vehicle manufacturing, Ford plans to produce a staggering 150, 000 units annually. This ambitious target equates to approximately 20% of the total F-series trucks sold by Ford in the previous year. As an all-electric iteration of America's best-selling truck, the F-150 Lightning aligns with the preferences and needs of existing Ford truck owners, eliminating the need to explore unfamiliar territories.

In its fundamental aspects, the F-150 Lightning maintains a semblance to its counterparts, with subtle differentiators like a charging port on one front fender and a faux charging port on the side. While additional lights adorn its front and rear, and the grille features a unique design, these distinctions may go unnoticed by casual observers. The absence of a front engine is a striking departure, replaced by a capacious storage space beneath the hood. This space serves not only for traditional storage but also hosts electrical plugs, transforming it into a functional power hub. During a Texas event, Ford demonstrated its versatility by filling the storage space with ice, converting it into a cooler for chilled beverages.

Powering the Lightning are two electric motors—one for the front wheels and another for the rear—resulting in standard full-time all-wheel-drive across all versions. From a utilitarian work truck with minimal amenities, offering a range of 230 miles, 426 horsepower, and 775 pound-feet of torque at a starting price of around $40, 000, to more luxurious models with six-digit price tags, the Lightning caters to a broad spectrum of users. Its performance, even with a 5, 000-pound trailer in tow, exemplifies its capabilities, with electric motors delivering instantaneous response, smooth acceleration, and a nearly silent operation.

The refined driving experience of the Lightning can be attributed to the improved weight balance achieved by strategically distributing heavy battery packs between the front and rear wheels. The adoption of independent rear suspension, distinct from the solid rear axle found in traditional F-150s, contributes to a more comfortable ride, especially on uneven terrains. The Lightning's electric motors, offering seamless power delivery without the need for gear shifts, enhance its reputation as a civilized highway cruiser.

Off-road prowess further distinguishes the Lightning, showcasing its capability to navigate steep, muddy trails effortlessly. The electric motors, providing precise and rapid power delivery to the wheels, excel in slippery conditions. Beyond traditional truck capabilities, the Lightning introduces novel features, including a substantial front trunk (frunk) with internal plugs and lights. Moreover, when connected to a home charger, it can serve as a backup power source during outages and power tools at a worksite.

While towing and hauling inevitably consume substantial electricity, impacting the truck's range, the Lightning offers commendable mileage on a single charge—230 to 320 miles, depending on the battery pack size. However, the challenge lies in the comparative advantage of traditional gas trucks with longer initial ranges and quicker refueling times. For most daily driving needs, the Lightning's range suffices, with overnight charging addressing any range concerns. The only setback for potential buyers lies in the high demand, as Ford has temporarily halted new orders due to overwhelming customer interest.

Rivian EV

The major distinguishing factor evident among customers of Rivian is the customer service experience which is highly endearing. A sentiment echoed by numerous Rivian owners who lauded both the company's customer service and the vehicle's quality, positioning their Rivians among the best they've ever owned.

Despite occasional flaws, such as a recent recall affecting a substantial number of Rivians, owners express satisfaction surpassing expectations for a new entrant in the automotive industry. A proactive approach from Rivian, reaching out to an owner who shared an issue on an online forum, reflects the company's commitment to addressing concerns.

In comparison to established players like Tesla and traditional automakers, Rivian, founded in 2009 and publicly listed in 2021, has garnered substantial attention and investments from industry giants like Ford and Amazon. Recognized by experts as a formidable contender among electric vehicle startups, Rivian has navigated challenges associated with simultaneous launches of three vehicles—the R1T, R1S SUV, and an Amazon delivery van.

However, the journey has not been without obstacles, including delivery delays, a significant drop in stock value, and a workforce reduction in July 2023. Owners, while acknowledging these hurdles, remain steadfast in their appreciation for the vehicles, which can carry a price tag of approximately $100, 000 based on selected options.

Testimonies from Rivian owners highlight the attention and admiration their R1Ts attract. Such encounters with curious onlookers in parking lots or fellow drivers capturing images while driving alongside are commonplace for Rivian owners.

Rivian's attentiveness extends beyond the initial purchase, with proactive monitoring of vehicles and swift responses to reported issues. Instances of indicator and warning light malfunctions or water ingress prompted direct communication from Rivian, assuring owners of prompt resolutions.

While the current level of service has garnered acclaim, some owners express reservations about the company's ability to maintain quality and service standards as production scales up. Rivian aims to produce 25, 000 vehicles in the current year, a substantial increase from the second quarter's delivery of fewer than 5, 000 vehicles. Questions about service quality and wait times loom as Rivian expands its production capacity.

Remarkably, despite opportunities to capitalize on the soaring demand for Rivians and earn substantial profits, owners convey an unwavering attachment to their vehicles. Ross Gale, a self-described business-oriented individual with minimal attachment to material possessions, exemplifies this sentiment. Despite a history of flipping cars for profit during the pandemic, he articulates a reluctance to part with his R1T, expressing disbelief at the prospect of selling such a prized possession.